WHEN DISASTER STRIKES!

The
Exxon Valdez
Oil Spill

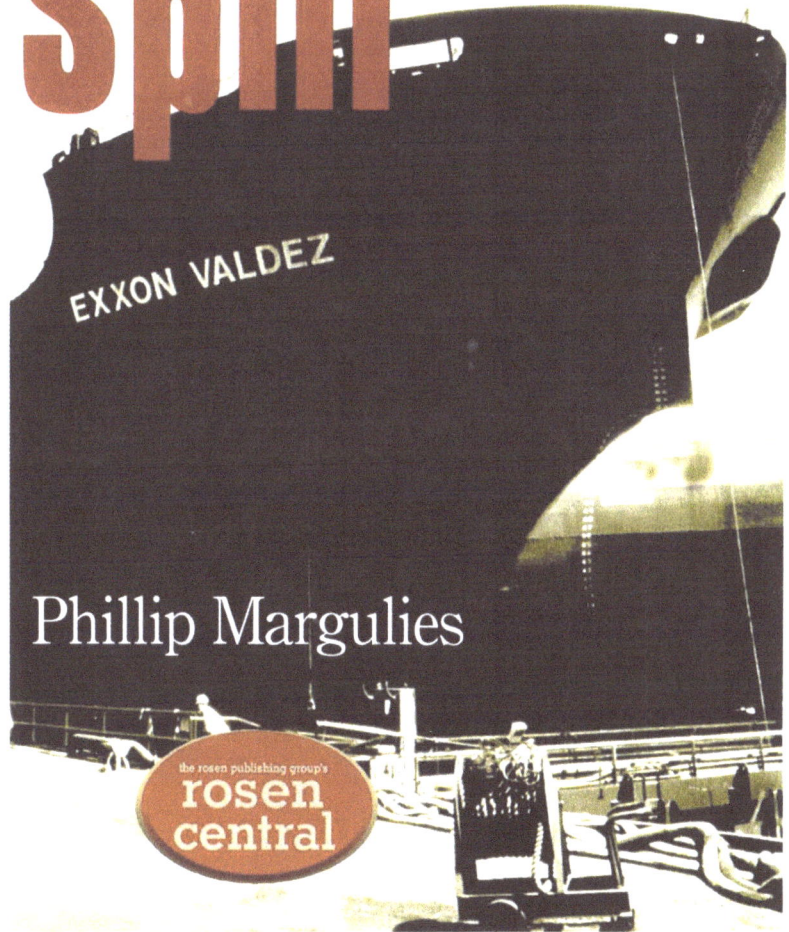

EXXON VALDEZ

Phillip Margulies

the rosen publishing group's
**rosen
central**

Published in 2003 by The Rosen Publishing Group, Inc.
29 East 21st Street, New York, NY 10010

Library of Congress Cataloging-in-Publication Data

Margulies, Phillip.
The *Exxon Valdez* oil spill / Phillip Margulies. — 1st ed.
p. cm. — (When disaster strikes!)
Summary: Discusses the factors and events that led to the 1989 *Exxon Valdez* oil spill in Prince William Sound, Alaska, the cleanup effort afterwards, and the long-term consequences of the disaster.
Includes bibliographical references and index.
ISBN 978-1-4358-8931-6
1. Oil spills—Environmental aspects—Alaska—Prince William Sound—Juvenile literature. 2. *Exxon Valdez* (Ship)—Juvenile literature. 3. Tankers—Accidents—Juvenile literature. [1. Oil spills—Alaska—Prince William Sound. 2. *Exxon Valdez* (Ship) 3. Tankers—Accidents.]
I. Title. II. When disaster strikes! (New York, N.Y.)
TD427.P4 M375 2003
363.738'2'097983—dc21

2001008523

Manufactured in the United States of America

On the cover and title page:
The *Exxon Valdez* sits in dry dock after the accident that spilled eleven million gallons of oil into Alaska's Prince William Sound.

Contents

The *Exxon Valdez* oil spill left tons of oil-soaked debris along the coastline of Alaska's Prince William Sound.

Introduction

Just after midnight on March 24, 1989, a giant oil tanker called the *Exxon Valdez* was making its way through Alaska's Prince William Sound, steering a tricky course between an ice floe and Bligh Reef. The ship was in trouble. Third Mate Gregory Cousins knew enough to be worried. A red buoy light in the distance told him the ship was off course. Cousins called the captain, Joseph Hazelwood, who was resting in his quarters. While they were discussing what to do next, the ship hit the reef. There was no explosion; the tanker ploughed slowly into the reef with only a few sharp shudders. The impact was enough for the crew to feel but not hard enough to knock anyone down.

The collision felt gentle, but the ship's side had been ruptured and its cargo was now leaking into the pristine sea. The crew soon saw the oil bubbling to the surface beyond the ship's edge. Within hours, the eyes of the world were focused on Prince William Sound, which had just suffered the worst oil spill in U.S. history.

Why did the *Exxon Valdez* oil spill happen, and what lessons did it teach us? To answer these questions, we first have to look at the role that oil plays in the modern world.

A Strategic Resource

During World War II, one of the basic strategies of the Allies (consisting of the United States, the United Kingdom, France, and Russia) and the Axis powers (Germany, Italy, and Japan) was to cut off each other's oil supplies. Without fuel, tanks and battleships are useless metal hulks, so control of oil meant the difference between winning and losing. Oil's importance to waging war successfully is the reason oil is often called a "strategic resource." At crucial moments—such as during war—access to oil can be essential to a nation's survival.

Like the war machine, the engines of peace also run on oil. Access to a plentiful supply of inexpensive oil is as much of a concern in peacetime as it is during a time of conflict. When the price of oil goes up, so does the price of almost everything else we buy, since so many things are made and brought to market with the help of oil.

OPEC Controls the Oil Supply

The industrialized countries of the West learned this hard economic lesson in the 1970s. At that time, the oil-exporting countries of the Arab world—nations that produced far more oil than they used—had united to form the Organization of Petroleum Exporting Countries (OPEC). Acting together, OPEC leaders could increase or reduce the amount of oil available in

OPEC's eleven member nations collectively supply about 40 percent of the world's oil output and possess more than three-quarters of the world's total known crude oil reserves.

the world and its price. To reduce the global supply of oil and increase prices, all the OPEC members had to do was agree among themselves to decrease the amount of oil it would produce and sell, which is exactly what OPEC did in the mid-1970s. The plan worked just as OPEC had hoped it would.

Since oil is a necessity, people simply paid more to get it, and oil and gas prices soared. So did the price of most other things, since oil and gas were needed to transport goods and power factories. As a result, the 1970s are remembered as a time of runaway inflation (when goods are scarce and prices are high) and energy shortages (Americans would line up for several hours on assigned days in order to get their gas tanks filled).

Politicians and voters in the United States became worried about the country's dependency on expensive oil from other, often hostile, countries. All over the world, governments and private companies scrambled to find new sources of oil, mostly by stepping up plans for drilling in places where oil had already been found.

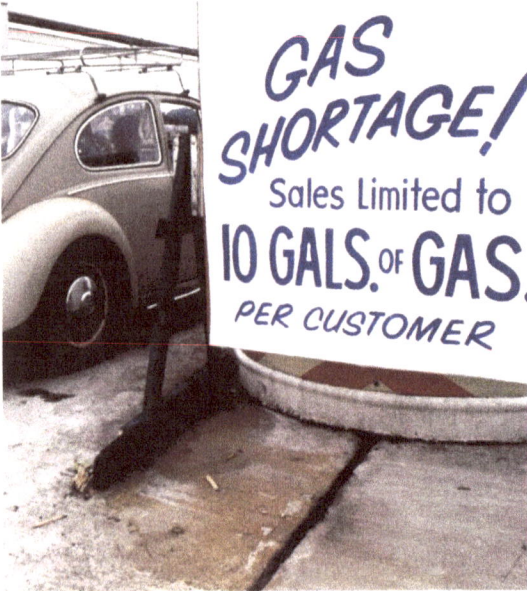

OPEC's oil embargo caused gasoline shortages throughout the United States during the 1970s.

High Prices Turn Oil into Black Gold

In addition to the political and strategic reasons some Western countries had for oil drilling, there was another incentive: profit. Every country and state in which oil could be found had a potential bonanza on its hands. With oil prices higher than ever, it was now worthwhile to look for oil in places where, a few years earlier, it would have been considered too expensive to drill.

Joseph J. Hazelwood: Captain of the *Exxon Valdez*

"Yeah, ah Valdez back, ah, we've, should be on your radar there, we've fetched up, ah, hard aground, north of Goose Island, off Bligh Reef, and, ah, evidently leaking some oil and we're gonna be here for awhile and, ah, if you want, ah, so you're notified, over."

Radio transmission from Hazelwood to the Coast Guard twenty minutes after the *Valdez* ran aground

The price of oil was so high in the 1970s, in fact, that oil company profits would also enrich federal, state, and local governments and private individuals. The government would get more money in taxes from the oil companies, and as a result tax rates could be lower for private citizens. Individuals would also benefit from the jobs that drilling and oil transport would create. In the case of Alaska, where oil reserves were rich and underdeveloped, a portion of the oil profits were promised to each citizen of the state in return for allowing companies to drill in Alaska's untouched environment. It was argued that drilling for oil at home would allow all Americans to win; everyone would be happy.

Not everyone, however, agreed that increased oil drilling was a good idea. Environmentalists expressed concerns about oil drilling. They were worried that an accident could rip oil tanks open, and the spilled oil would kill wildlife on a vast scale.

This was exactly the accident waiting to happen in Prince William Sound in the early morning of March 24, 1989.

Warnings Ignored

Though *Exxon Valdez* captain Joseph Hazelwood would receive the brunt of the blame for the accident in Prince William Sound that led to one of the worst environmental disasters in history, he was the last link in a long chain of errors. The *Exxon Valdez* oil spill was a disaster born of misleading claims, false promises, mismanagement, and greed, and the blame could be extended far and wide. The seeds of the accident were planted a decade earlier, in the midst of the nation's energy crisis.

The Trans-Alaska Pipeline

Two conflicting interests—tapping oil reserves at home and protecting the environment—went head to head in 1973 when seven oil companies announced their development proposal for the oil reserves of Alaska's North Slope. The plan was this: They would build a pipeline through Alaska, from the North Slope oil fields to Prince William

The trans-Alaska pipeline links oil wells in Prudhoe Bay, Alaska, with the port of Valdez. The pipeline transports 17 percent of the United States's total oil production.

Sound. From there, the oil would be moved by tankers—large cargo ships designed especially for carrying liquids in bulk—south along the Pacific coast of Canada to the continental United States.

The pipeline had to be approved by the U.S. Congress, and there were many arguments over what should be done. Some environmentalists were against both the drilling and the pipeline. Others said that the oil companies should instead build a longer pipeline, one that would go south through Canada. That way the tankers would

Thursday, March 23

9:26 PM

After loading has finished, the *Exxon Valdez* leaves the Alyeska terminal in Port Valdez, bound for Long Beach, California.

not have to take the oil through the waters of Prince William Sound. They pointed out that Prince William Sound, with its ice floes, narrow passages, and poor visibility, was difficult to navigate, making accidents and oil spills more likely. The longer pipeline would also have cost more to build, however, making it an unattractive option for the oil companies and Congress. As it was, the shorter trans-Alaska pipeline from the North Slope to Prince William Sound was the most expensive private construction project ever up to that time.

In the end, the project's sponsors convinced politicians and voters that the risk of oil spills was low. The technology for preventing oil spills was more advanced than ever, they said, as was the technology for limiting the damage if a spill ever occurred. The oil companies presented a detailed plan that explained what could be done if a spill occurred. This spill contingency plan was written into law as part of a deal struck between the U.S. Congress and the representatives of the seven oil companies to permit oil drilling in the North Slope of Alaska.

So, the North Slope oil fields were developed, and drilling yielded vast quantities of crude oil. The trans-Alaska pipeline was built, and the giant tankers began moving the oil through the narrow and treacherous straits of Prince William Sound south to the refineries on the West Coast of the United States.

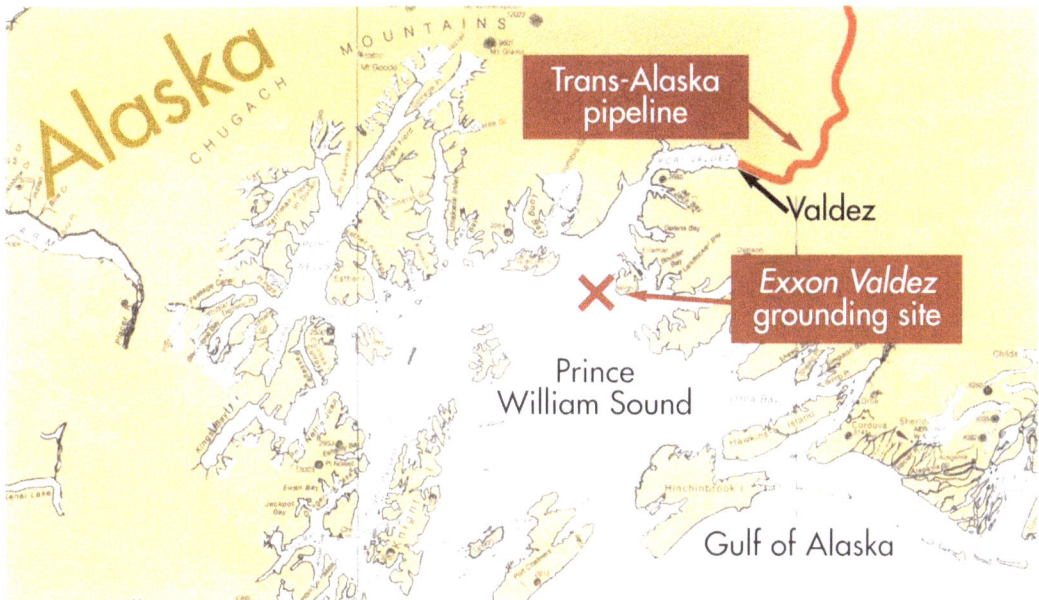

Trans-Alaska pipeline

Valdez

Exxon Valdez grounding site

Prince William Sound

Gulf of Alaska

Fifty-six days after the *Exxon Valdez* ran aground on Bligh Reef, its eleven million gallons of spilled oil had drifted 470 miles southwest, polluting large stretches of Prince William Sound and the coast of the Alaskan peninsula.

Success Brings Lower Oil Prices

Ironically, the very success of the trans-Alaska pipeline helped to make an oil spill more likely. Drilling in Alaska resulted in vast new sources of oil. At the same time, OPEC had once again boosted oil production. By the 1980s, talk of oil shortages had turned to talk of an oil "glut," which is an excessively large supply. The price of oil went down, sales of big, gas-guzzling cars picked up, and people began to take oil for granted once again. One group of people not delighted by the abundance of cheap oil were those employed in the

Thursday, March 23

10:17 PM

The *Exxon Valdez* enters the Valdez Narrows, the most dangerous leg of the journey.

oil industry. A boon to consumers, the oil glut represented nothing but hard times for the oil companies. Low oil prices put pressure on everyone involved in transporting oil to cut corners. Safety measures that had been taken to win approval for the pipeline fell by the wayside and were often ignored.

To run the trans-Alaska pipeline and transport oil safely to market, a consortium of several oil companies—including Exxon, Hess, British Petroleum, and Unocal—formed a joint company named Alyeska. It was Alyeska's responsibility to minimize pollution caused by the pipeline and to be prepared to respond quickly and effectively to any oil spills. Alyeska officials promised that it could clean up as many as 100,000 barrels of oil in seventy-two hours, if necessary. But what if there was a bigger spill? The Alaska Department of Environmental Conservation (ADEC) asked this very question in 1986. At ADEC's request, Alyeska came up with a contingency plan for a 200,000 barrel spill. It assured ADEC that the odds that such a spill would happen were only once in 241 years.

No Punishment for Rule-Breaking

Alaska's state government was supposed to be a watchdog over Alyeska, to make sure that Alyeska was carefully monitoring oil pumping and transport operations and was prepared to react to an oil spill. However, the state needed its share of the money that oil drilling generated. Eighty percent of Alaska's revenue comes from the oil business. Even individual Alaskans had a stake in oil company profits. A portion of the income generated by Alaskan oil sales went into a special fund, from which each individual Alaskan received a dividend check every year. In addition, in return for agreeing to the North Slope drilling, residents of Alaska paid no state income tax. Oil profits were driving Alaska's economy and affecting the decisions of businesspeople, politicians, bureaucrats, and regular taxpayers.

The fact that Alaskans reaped so many financial benefits from drilling made it difficult for the state to enforce its own laws. Cracking down on Alyeska would mean smaller profits for everyone involved. Again and again, state inspectors found that Alyeska was not complying with the operational and environmental safety rules and not keeping its promise to be prepared for a large oil spill. Yet Alyeska was never seriously punished for its broken promises.

George Nelson: President of Alyeska

"The response was not late. Our tug was out there in two hours . . . We were late with the booming equipment and skimming, but with this type of spill, that's not what the response is . . . You cannot suck up 240,000 barrels of oil. The mechanical equipment basically just doesn't exist . . . We still believe we have had an outstanding record."

**From reporting in the *Anchorage Daily News*,
March 28, 1989**

The Grounding of the Exxon Valdez

On March 22, 1989, the *Exxon Valdez* docked at the Alyeska terminal. The next day it took on its cargo of 1,286,738 barrels of crude oil. The process began early in the morning and lasted until after nightfall.

The captain, Joseph Hazelwood, was ashore for most of that time. Some witnesses later remembered seeing him consume several alcoholic drinks in the course of the day.

By 9 PM, the ship was ready to set sail and begin its usual winding way through the Valdez Narrows, considered the riskiest part of the trip out of Prince William Sound.

As insurance, escort tugs accompany ships through the narrows. In addition, while ships are in the narrows, a harbor pilot—specially trained in the hazards of local waters—is always on board until the ship reaches the safety of open water. On this day, two escort tugs and a harbor pilot named Ed Murphy helped usher the *Valdez* through the narrows (though one tug dropped out early on).

Off Course

By 11:24 PM, the tanker was safely out of the Valdez Narrows, ready to begin what was considered a less dangerous part of the journey out of Prince William Sound. His job successfully completed, Ed Murphy sailed away from the tanker in a pilot boat. Captain Hazelwood sent a radio message to Traffic Valdez, the Coast Guard's radio monitoring station. He told the Coast Guard that he planned to take the tanker the rest of the way out of Prince William Sound via an "inbound lane"—a path usually used for tankers entering the sound, not leaving it. Captain Hazelwood explained that he was taking this step to avoid the ice his ship's radar had sighted in the outbound lane. Traffic Valdez gave him the schedules of tankers coming toward Hazelwood's ship. There would be plenty of time for the *Exxon Valdez* to use the inbound lane on its way out of the sound. In a later radio message, Captain Hazelwood told Traffic Valdez that he planned to reduce speed "to wind my way through the ice."

Captain Hazelwood did not reduce speed, however, as he had told the Coast Guard he would. Nor did he stick to the course he said he would follow. Instead, he took the *Exxon Valdez* much farther off its

Thursday, March 23

11:31 PM

Captain Hazelwood radios his plans to avoid ice by slowing down and changing his course.

usual path than he had earlier proposed. The tanker was suddenly faced with two dangers: Bligh Reef on one side and an ice floe on the other. At this point, Captain Hazelwood had three choices. He could have stopped the ship and waited until the ice floe moved, or he could have slowed down to wind his way though the ice as he had told Traffic Valdez he would. Instead he made a reckless third choice, which was to order that the ship make a daring turn and enter a narrow gap between the ice and the reef.

The *Exxon Baton Rouge*, the smaller ship, attempts to offload the unspilled crude oil from the *Exxon Valdez* on March 26,1989.

Collision

At 11:53 PM on March 23, after giving orders to execute this tricky maneuver, Captain Hazelwood left the bridge, leaving Third Mate Gregory Cousins in charge. Cousins continued to communicate with Hazelwood by ship phone. "I think there's a chance we may get into the edge of the ice," he told the captain at 11:55 PM. A few minutes later he called again, saying, "I think we are in serious trouble."

At 12:04 AM on March 24, while Cousins was on the phone with Captain Hazelwood, the ship hit Bligh Reef with a shudder. Captain Hazelwood ran immediately to the bridge. The air was suddenly filled with the smell of crude oil. One crewmember saw oil shooting up forty or fifty feet into the air. Clearly, the ship's hull had been pierced. Gauges in the control room confirmed the crew's worst fears: Eight of the eleven cargo tanks were ruptured, and the tanker was leaking oil fast.

"You're Looking at It"

At 12:40 AM, Captain Hazelwood began giving a series of commands intended to prevent the ship from capsizing and harming the people on it. These measures eventually helped stabilize the ship.

Soon after 3 AM, three men boarded the ship. One of them was Chief Warrant Officer Mark Delozier, who found Captain Hazelwood drinking coffee. Delozier said later that he smelled alcohol on Captain Hazelwood's breath but that his speech and behavior did not indicate that he was in any way impaired or drunk. He asked Hazelwood what had caused the accident, and the captain said, "You're looking at it."

Thursday, March 23

11:36 PM

Hazelwood changes course again, bringing the *Exxon Valdez* close to the treacherous Bligh Reef.

By 6 AM, an oil slick three miles long and two miles wide was drifting south from the tanker. Ultimately, the ship would leak eleven million gallons (about 250,000 barrels) of oil into the water. It was enough to cover the area of a football field thirty feet deep in crude oil. Roughly four-fifths of the ship's cargo of oil did not leak into the sea and was siphoned out of the *Valdez* into three tankers involved in the salvage and cleanup operation.

Joseph Hazelwood at his trial in 1990

Captain Joseph Hazelwood

Captain Joseph Hazelwood was forty-two years old at the time of the grounding of the *Exxon Valdez*, and he had made well over 100 round trips through Prince William Sound. He had a good reputation as a captain. He had often shown himself to be brave and resourceful in a crisis. In 1987 and 1988, Exxon gave the *Exxon Valdez*, under Captain Hazelwood's command, an award for safety and performance. Coworkers said that he liked to drink but that he always knew when to stop.

Job performance and drinking habits can change over time, however. Captain Hazelwood was arrested twice for drunk driving in the 1980s—the second time in 1989. In 1985, an Exxon manager wrote in a memo that Hazelwood had "violated company alcohol policy on at least several occasions." It was never clearly established whether or not Hazelwood was drunk on the night of the accident, and none of the charges later brought against him were alcohol-related.

Rescuers

Many of the steps taken to limit the damage caused by the *Exxon Valdez* grounding ended up only making matters worse. In a way, they served to prolong the disaster and deepen its effects. The decisions made after the spill (and the long delays during which no decisions were made) had large consequences for the wildlife of the spill region, all of them bad. Like nearly everything else connected with the *Exxon Valdez* spill, the actions taken after the accident are mired in controversy.

Who should have led the response to this emergency? If preparation for a crisis had been adequate, the answer should have been absolutely clear. It was far from clear, however, who was in charge. Precious time was lost as the buck was passed among Alyeska, the Coast Guard, Exxon, the state of Alaska, and the U.S. federal government. No one seemed interested in accepting responsibility for containing and cleaning up the spreading mess in Prince William Sound. Complicating any attempt at a cleanup was the remoteness of the area, which was reachable only by helicopter and boat.

Alyeska Unprepared

Alyeska was supposed to handle the initial stages of cleanup in the event of a spill. It was soon clear, however, that Alyeska was not up to the job. Its plan for spills of this magnitude (the possibility of which had never really been taken seriously) stated that an emergency response would be under way within two to five hours. Instead, Alyeska first arrived on the scene fifteen hours after the spill, and by then could do little to contain it. Much of the necessary oil containment equipment was buried in warehouses under heaps of other equipment. The oil barge that was supposed to contain the oil that Alyeska collected from the spill was unavailable. Alyeska put skimmers (oil-collecting boats) on the scene twenty-four hours after the accident, but their work soon stopped because of the absence of the collection barge. They could gather no more oil until they unloaded the quantities they had already collected onto a barge.

Friday, March 24

12:04 AM

The *Valdez* runs aground on Bligh Reef, gashing its hull and leaking hundreds of thousands of gallons of oil.

Disaster Area?

The governor of Alaska called on President George Bush to declare the spill a national disaster. This would have allowed the U.S. Congress to provide money to help control and clean up the spill. At this point, the U.S. federal government would have taken charge of the cleanup. However, the Bush administration refused to declare the spill a national disaster. The thinking was that the spill was Exxon's fault; Exxon should take care of it, not American taxpayers.

Oil containment booms protect the Armin Koernig Salmon Hatchery at Sawmill Bay after the *Exxon Valdez* oil spill. Seaplanes were employed to assist in cleaning up the eleven million gallons of spilled oil.

Exxon Takes Charge

Even before the Bush administration came to its decision, executives from Exxon announced that the company would take over the cleanup from Alyeska. Exxon was not really any better equipped to clean up the oil that had already spilled into Prince William Sound than Alyeska was, however. Though oil spills are quite frequent events, Exxon responded to this one as if it was the first the company had ever encountered. Still more time was lost as Exxon executives and other agencies argued over whether to clean up the oil with booms and skimmers, with chemical dispersants, or by burning or hosing it away. Much time was wasted as the company studied and considered how to handle the ever-worsening situation. The time for deliberating had passed, yet inaction prevailed and the oil slick grew larger and larger, threatening the Prince William Sound fishery, miles of shore-line, ten million migratory shore birds, and hundreds of sea otters, harbor porpoises, sea lions, and whales.

Skim It, Burn It, Hose It, or Disperse It?

Each of the many ways to deal with oil spilled in the ocean has draw-backs. With skimming, or "containment and recovery," oil is sur-rounded with booms (floating devices that deflect and corral oil) and collected with skimmers. A lot of the oil escapes and must be dealt with by other methods. To be effective, skimming has to be done immediately, before the oil has a chance to expand over vast areas of

Orange and yellow oil containment booms stretch across the water during the cleanup of Prince William Sound.

the ocean, carried far and wide by the tides, and before it starts to pollute the beaches. Skimming may have been the best option, but Alyeska's poor preparation and delayed response made it impractical for the *Exxon Valdez* spill. Skimmers only became available twenty-four hours after the accident. At that point the spill was spreading and was no longer containable. Bad weather further slowed the oil recovery efforts, as did the oil and kelp that continuously clogged the equipment, but skimming remained the chief containment strategy throughout the cleanup.

Burning spilt oil turns it into a gas that disperses in the atmosphere. The oil industry claims that burning the oil adds no more to the world's total amount of air pollution than if the oil had been burned as fuel. While this is true, the problem is that the burning of thousands of gallons of fuel happens in one small area—Prince William Sound, in this case—rather than in thousands of cars, buildings, and homes spread out around the entire country. Exxon experimented with burning the oil off in the early stages of the spill but decided against this course in the end, fearing that a fire of this scale would have been deadly to the region's fish and wildlife.

Dispersants act like dishwashing detergent, breaking up the oil into more easily collectable clumps. The main drawback to using chemical dispersants to clean up an oil spill is that toxic residues from the oil (that might otherwise evaporate if burned off) stay in the water and are consumed by fish that human beings may later eat. As it turns out, however, Exxon could not use dispersants on a large scale anyway because they simply were not available in large enough quantities. Like Alyeska, Exxon had not been prepared for a spill of this size. In addition, dispersants must be activated by wave action; the waters of Prince William Sound were initially very calm, and Exxon engineers concluded that the dispersants were not working. Their use was quickly discontinued.

A fourth method of oil removal, high-pressure hosing using hot water and chemicals, was widely used to clean up the shorelines. The oil that was not trapped and collected by the booms and skimmers (which was the bulk of the spill, unfortunately) spread out to sea or washed ashore on the previously unspoiled beaches of the Alaskan coast. High-pressure hosing cleaned beaches of oil, but it also killed every living thing on them. Today, the untreated beaches are in better shape ecologically than the treated ones.

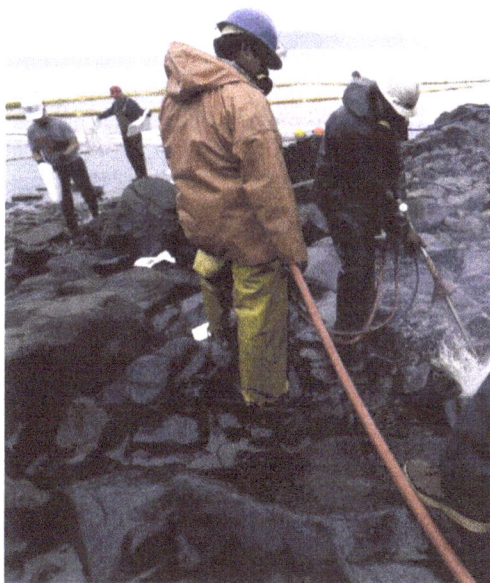

Workers use chemical sprays to try to clean the oil off rocks in Prince William Sound after the *Exxon Valdez* spill.

Friday, March 24

4:01 AM

Alyeska issues an official oil spill notification, triggering its emergency response procedures. The message reads, "This is not a drill."

The Spill Spreads

While the various agencies and companies argued over who was in charge and what should be done, almost eleven million gallons of oil spilled into Prince William Sound. By the night of March 26, two days after the spill, the slick covered an area of 18 square miles. During this period, the weather was unusually calm. The relatively unruffled seas made conditions as ideal as they ever would be for a cleanup. On March 27, however, this good luck failed as a storm drove the oil slick south and west. By April 4, eleven days after the spill, the slick covered an area of 1,000 square miles. Scientists believe that 40 percent of the spill was eventually washed up on beaches, 35 percent of it evaporated, and 25 percent entered the Gulf of Alaska, either washing ashore or out to sea. Some 790 miles of shoreline within Prince William Sound had been oiled, a quarter of that heavily so. If there was a silver lining, it was that the spill could have been more catastrophic. Roughly 42 million gallons of oil did not leak and were safely removed from the damaged ship's tanks.

Wildlife Threatened

Though the most environmentally sensitive areas of the sound, such as seal pup locations and fish hatcheries, were targeted for the earliest cleanup efforts, rescue efforts were slow and animals all

over the spill region were begin-
ning to sicken and die from
many different causes. Some
wildlife died after breathing poi-
sons that the oil sent into the air
and from drinking the oil-tainted
water. Many other animals died
from exposure to the cold
Alaskan air; oil that coated their
fur, scales, or feathers made
them unable to regulate their
body temperatures.

Alaska's Department of the
Interior ordered Exxon to set up
animal rescue centers to try to
save and treat as many of the
affected animals as possible.
Exxon hired professionals to
manage them. Around half the animals taken to the centers soon
died despite treatment. Sadly, for one reason or another, most of the
mammals and nearly all of the birds that were treated at the rescue
centers did not survive the year.

Workers with the International Bird Rescue Project clean an oil-soaked loon at the bird rescue center in Valdez, a week after the *Exxon Valdez* ran aground.

Animals of hundreds of species were killed in the *Exxon
Valdez* oil spill. Twenty-three species had their populations reduced
enough to make it doubtful they would return to their former
numbers. About 250,000 seabirds were killed, along with 2,800 sea
otters, 1,000 cormorants, and hundreds of loons, harbor seals, and
bald eagles.

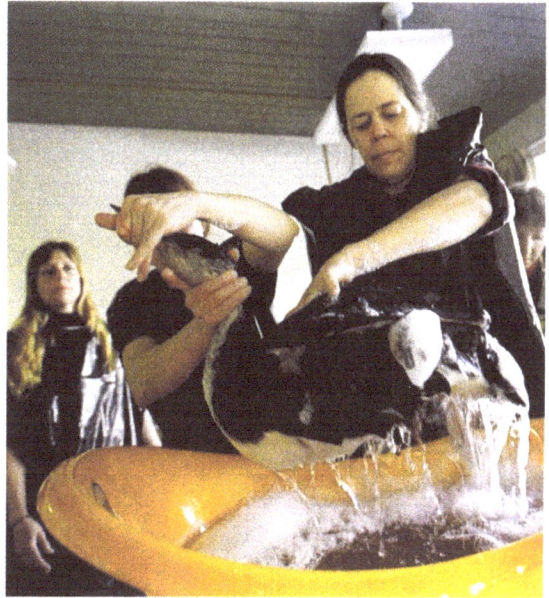

Fishermen

Other than oil, the major industry operating in Prince William Sound was fishing. The fishermen were never happy about the growth of the oil business in their neighborhood. They had dreaded the effects of an oil spill that could potentially destroy their catch, and many of them were sure an accident would happen sooner or later. Through their organization, Cordova District Fishermen United (CDFU), they had fought to have the trans-Alaska pipeline extended through Canada to the lower forty-eight states, in order to keep the oil tankers out of their waters. The *Exxon Valdez* oil spill confirmed their worst fears—it ruined the fishing not only in Prince William Sound but also in much of the Gulf of Alaska for years to come.

A fisherman removes oil-soaked debris from the water near Eleanor Island, Alaska, after the *Exxon Valdez* fouled the area.

Henry Milette: Fisherman

"It's gonna affect the bottomfish, the shrimp, the snapper. That's how I make my living. Well, I guess we're professional oil cleaners, now."

From reporting in the *Anchorage Daily News*, March 27, 1989

Alaska's Native Americans

Alaska's many Native American tribes are not all alike in their attitude toward technology. Some embrace it, while others lead a traditional way of life based on hunting and fishing the wildlife of the region. The inhabitants of the village of Tatilek are native Alaskans dedicated to older traditions. Most of the people of Tatilek had lived in cities at some point but had rejected the urban life, choosing instead to return to the ancient daily customs of their ancestors. Hunting and fishing are central to this centuries-old way of life. The *Exxon Valdez* oil spill threatened the Tatilek by killing the animals upon which they rely.

4

The Long Aftermath

The effects of the *Exxon Valdez* oil spill are still being felt today, as the land, water, and wildlife of Prince William Sound continue to try to rebound from the environmental disaster, and the oil companies—Exxon in particular—continue to try to recover from the public relations disaster. Meanwhile, one man—the captain of the *Valdez*—seeks to rebuild his life and career and disappear from the public eye.

Culprits

Captain Joseph Hazelwood lost his job with Exxon, was stripped of his ship captain's license, and spent several years fighting criminal and civil charges against him. He was convicted of only one charge: negligent discharge of oil. He was sentenced to 1,000 hours of community service in Alaska, spread over several years. Hazelwood regained his captain's license but has been unable to find work as a tanker captain. He has taken other seafaring jobs, however, such as lobster fishing. He is currently employed in New York as a claims adjuster for the maritime lawyers who defended him.

Hazelwood makes breakfast at a cafe in Anchorage as part of his community service sentence.

The ship, the *Exxon Valdez*, has been renamed the *Exxon Mediterranean*. Barred by law from Prince William Sound, it carries crude oil from the Middle East to Europe.

Exxon was found guilty of having committed an "environmental crime." Between the costs of the cleanup, criminal penalties, and out-of-court settlements, the oil company has spent more than $3 billion to try to make things right. In 1994, as a result of a separate class-action lawsuit filed by fishermen, natives, and property

Friday, March 24

2:30 PM

An Alyeska barge containing booms and skimmers arrives at the accident scene, 14 ½ hours after the massive oil spill began.

owners hurt by the spill, a jury ordered Exxon to pay an additional $5 billion. Exxon refuses to pay this penalty and has so far avoided doing so. In November 2001, an appeals court struck down the $5 billion penalty, judging it to be excessive, and ordered the case back to trial court to determine a smaller fine.

Attorney General Dick Thornburgh announces the fines against Exxon at a press conference in Washington, D.C., on March 13, 1991. Exxon was ordered to pay $900 million to clean up the *Exxon Valdez* oil spill and $100 million in criminal fines.

Victims

Scientists are still studying the long-term effects of the oil spill on Prince William Sound. Today, many of the beaches again look beautiful, but some of the oil spilled in 1989 can still be found up to 400 miles from the place where the *Exxon Valdez* ran aground. Beneath the shore's rocks, sand is covered with oil that has hardened like asphalt. On warm days, the smell of oil is still carried on the wind. The populations of twenty-three species of birds, fish, and marine animals were seriously reduced by the spill. Of these, only two—the bald eagle and the river otter—have recovered to their pre-1989 population levels.

The effects of the spill on wildlife are widely disputed, though. Animal populations change all the time as a result of many different things, and it is hard to prove that any one event causes the numbers of a particular species to decline. Due to this uncertainty, environmentalists and oil industry spokespeople tell different stories, and each side can find scientists to support its viewpoint.

Preventing Spills

Between 1989 and 1990, Alaska's state legislature passed a dozen new laws addressing oil spill prevention, response, and oversight in the wake of the *Exxon Valdez* disaster. At the federal level, the U.S. Congress passed the Oil Pollution Act in 1990. This new law ordered changes to be made that would help prevent future oil spills and reduce damage to the environment when spills occurred. One of the most important new rules gives the oil industry until 2015 to switch to double-hulled tankers. In tankers of this kind, a second, inner hull continues to contain the oil even when the outer hull is damaged. The use of a double-hulled design would reduce the risk of oil spills by 75 percent. The *Exxon Valdez* was only a single-hulled tanker.

The oil companies have been slow to make the transition to double-hulled tankers. Ten years after the spill, only around one-fourth of the tankers in operation were double-hulled, and only three double-hulled ships were on order. Oil industry spokespeople now say they might not meet the government's deadline.

The *Arco Endeavour* is a double-hulled tanker designed to comply with post–*Exxon Valdez* oil spill laws.

The U.S. Coast Guard's ability to monitor traffic in Prince William Sound has improved since 1989. New radar systems provide better tracking abilities in difficult weather conditions and integrate ground radar with additional information provided by satellites. Today, if a tanker in Prince William Sound were to go as far off course as the *Exxon Valdez* did, the Coast Guard would find out immediately and advise the ship's crew of their error and how to correct it. While the switch to double-hulled tankers has yet to occur, these traffic control improvements are currently in place and greatly reduce the risk of accidents that would result in devastating oil spills.

Prevention Isn't Just Technology

The most important changes that have been introduced to prevent future *Exxon Valdez*–like disasters are not engineering improvements. Instead, they are just sensible precautions that demonstrate a greater willingness, on the part of the government and the oil industry, to spend money on safety. For example, the U.S. Coast Guard now monitors fully laden tankers from the moment they leave port until they safely clear Prince William Sound; before, the Coast Guard monitored the ships for only part of this journey. In addition, two escort vessels now accompany each tanker while passing through the entire sound (there was one escort vessel with the *Exxon Valdez*). In addition, specially trained marine pilots with extensive experience in Prince William Sound are now aboard the ship during its entire voyage through the sound, not just through Valdez Narrows, as was the case with the *Exxon Valdez.*

There have also been many improvements in planning for another spill. Luckily, Prince William Sound has not seen another spill on the scale of the *Exxon Valdez*. But until there is one, we won't know whether improvements in planning will make a difference in containing the damage. On average, a total of 100 million gallons of oil spills each year.

Conclusion

From any perspective, the *Exxon Valdez* oil spill counts as an important episode in the history of human shortsightedness, since most safety precautions being taken now to prevent another such disaster could just as easily have been in place before 1989 if the decision makers had not been so concerned with short-term profits.

It is said that an ounce of prevention is worth a pound of cure. This sounds like a good bargain. In reality, though, prevention costs money.

A group of protestors march through the streets of Anchorage, Alaska, on March 24, 1999, in observance of the ten-year anniversary of the *Exxon Valdez* oil spill in Prince William Sound.

To minimize the oil industry's damage to the environment, ordinary people have to be willing to pay higher prices for gas and electricity. Through their representatives in government, they have to insist that laws be enforced. They have to insist that the penalties be steep enough to inflict real financial damage on a company that breaks the rules, so that in the long run it is cheaper for a company to comply with the law than to violate it.

Above all, the public has to show that it has a longer attention span than the cynics say it has. We have to prove that we continue to care about creating a balance between corporate profits and the well-being of the environment, years after the images of oil-slicked sea otters have disappeared from the newspapers and television screens.

Glossary

ADEC   Alaska Department of Environmental Conservation.

barge   A roomy, flat-bottomed boat used chiefly for the transport of goods, usually propelled by towing.

bow   The forward part of a ship.

buoy   A floating object moored to the bottom of a body of water to mark something lying under the water.

class-action lawsuit   A legal action brought by a large group of people with a common complaint.

crude oil   Unrefined petroleum; oil in its natural, unprocessed state.

dispersants   Chemicals that help break down one substance to promote its distribution within another substance (for example, oil in water).

escort tug   A tugboat used or designed to assist a larger ship in navigation.

hull   The frame or body of a ship.

ice floe   A usually large, flat, free mass of ice floating in the sea.

narrows   A strait, or smaller body of water, connecting two larger bodies of water.

national disaster An official emergency for which the United States federal government can provide special assistance, funded by the federal budget.

OPEC Organization of Petroleum Exporting Countries.

private enterprise Private, profit-making organizations, in contrast to government agencies funded by taxpayers.

radar A system consisting of a synchronized radar transmitter and receiver that emits radio waves. When these radio waves hit a solid object, a reflection is created that allows a radar operator to track the object's movements.

skimmers Boats designed to collect ("skim") oil from the surface of the water.

sound A bay or recess in the shore of a sea, lake, or river.

strategic resource Material required for the conduct of war but not available in sufficient quantities in one's own country.

tanker A cargo ship fitted with tanks for carrying liquid in bulk.

toxic Poisonous.

trans-Alaska pipeline A pipeline built to carry oil across the state of Alaska, from the drilling fields near the North Slope's Prudhoe Bay to a terminal in the town of Valdez, near Prince William Sound.

tug Short for tugboat; a strongly built, powerful boat used for towing.

For More Information

Organizations

Alaska Department of Environmental Conservation (ADEC)
Anchorage Office
555 Cordova
Anchorage, AK 99501
(907) 269-7500
Web site: http://www.state.ak.us/local/akpages/ENV.CONSERV/
    dspar/dec_dspr.htm

Alaska SeaLife Center
P.O. Box 1329
Seward, AK 99664
(800) 224-2525
Web site: http://www.alaskasealife.org

Alyeska Pipeline Service Company
1835 South Bragaw Street, MS-542
Anchorage, AL 99512
(907) 834-6620
Web site: http://www.alyeska-pipe.com

National Oceanic & Atmospheric Administration (NOAA)
Office of Response and Restoration
1305 East-West Highway
Silver Spring, MD 20910
(301) 713-2989
Web site: http://www.noaa.com

Web Sites

Due to the changing nature of Internet links, the Rosen Publishing Group, Inc., has developed an online list of Web sites related to the subject of this book. This site is updated regularly. Please use this link to access the list:

http://www.rosenlinks.com/wds/exva/

For Further Reading

Blashfield, Jean F., and Wallace B. Black. *Oil Spills.* Chicago: Children's Press, 1991.

Brown, A. S. *Fuel Resources.* Toronto, ON: Franklin Watts, 1985.

Doherty, Craig A., and Catherine M. Doherty. *The Alaska Pipeline.* Woodbridge, CT: Blackbirch Press, 1997.

Lampton, Christopher. *Oil Spill.* Brookfield, CT: Millbrook Press, 1992.

Markle, Sandra. *After the Spill.* New York: Walker & Co., 1999.

Pringle, Laurence P. *The Environmental Movement: From Its Roots to the Challenges of a New Century.* New York: HarperCollins, 2000.

Siy, Alexandra. *Arctic National Wildlife Refuge.* Toronto, ON: Maxwell MacMillan Canada, 1991.

Smith, Roland. *Sea Otter Rescue: The Aftermath of an Oil Spill.* New York: Cobble Hill Books/Dutton, 1990.

Stefoff, Rebecca. *Environmental Disaster.* New York: Chelsea House Publishers, 1994.

Bibliography

Anchorage Daily News. "Hard Aground: Disaster in Prince William Sound." 1989–1999. Retrieved October 2001 (http://www.adn.com/evos/index.html).

Clark, Maureen/ABCNews.com. "*Valdez* Oil Spill's Stain Lingers." ABCNEWS.com. 1999. Retrieved October 2001 (http://abcnews.go.com/sections/science/DailyNews/exxonvaldez990322_main.html).

Fingas, Merv, ed. *The Basics of Oil Spill Cleanup*. Boca Raton, FL: CRC Press, 2000.

Hayes, Miles O. *Black Tides*. Austin, TX: University of Texas Press, 2000.

Keeble, John. *Out of the Channel: The* Exxon Valdez *Oil Spill in Prince William Sound*. Seattle, WA: University of Washington Press, 1999.

Lord, Nancy. *Darkened Waters: A Review of the History, Science, and Technology Associated with the* Exxon Valdez *Oil Spill and Cleanup*. Homer, AK: Homer Society of Natural History, 1992.

Loughlin, Thomas R., ed. *Marine Mammals and the* Exxon Valdez. San Diego, CA: Academic Press, 1994.

Markle, Sandra. *After the Spill: The* Exxon Valdez *Disaster, Then and Now*. New York: Walker and Co., 1999.

Prevention, Response, and Oversight Five Years After the Exxon Valdez *Oil Spill*. Fairbanks, AK: University of Alaska Sea Grant, 1995.

Think Quest. "Prince William Sound: Paradise Lost?" 1997. Retrieved October 2001 (http://library.thinkquest.org/10867/home.shtml).

United States Environmental Protection Agency. "*Exxon Valdez*." March 1999. Retrieved October 2001 (http://www.epa.gov/oilspill/exxon.htm).

Index

About the Author

Phillip Margulies is a freelance writer living and working in New York City.

Photo Credits

Cover, p. 1 © Brent Clingman/Timepix; pp. 4–5, 27, 30 © Natalie Fobes/Corbis; p. 7 © Bettmann/Corbis; p. 8 © Owen Franken/Corbis; p. 11 © Dave Bartruff/Corbis; p. 13 courtesy of NOAA; pp. 18, 33, 34, 36, 39 © AP/Wide World Photos; pp. 20, 29 © AFP/Corbis; pp. 24, 26 © Roy Corral/Corbis.

Series Design and Layout

Les Kanturek

www.ingramcontent.com/pod-product-compliance
Lightning Source LLC
Chambersburg PA
CBHW050911210326
41597CB00002B/95